DISASTROUS DEATHS

DEATH BY AWFUL ACCIDENTS

by
Mignonne
Gunasekara

BEARPORT
PUBLISHING

Minneapolis, Minnesota

Library of Congress Cataloging-in-Publication Data

Names: Gunasekara, Mignonne, author.
Title: Death by awful accidents / by Mignonne Gunasekara.
Description: Minneapolis, MN : Bearport Publishing Company, [2022] |
Series: Disastrous deaths | Includes bibliographical references and index.
Identifiers: LCCN 2020058665 (print) | LCCN 2020058666 (ebook) | ISBN 9781636911670 (library binding) | ISBN 9781636911724 (ebook)
Subjects: LCSH: Accidents–Miscellanea–Juvenile literature. | Violent deaths–Miscellanea–Juvenile literature. | Biography–Miscellanea–Juvenile literature. | Death–Miscellanea–Juvenile literature.
Classification: LCC HV675.5 .G86 2022 (print) | LCC HV675.5 (ebook) | DDC 363.1–dc23
LC record available at https://lccn.loc.gov/2020058665
LC ebook record available at https://lccn.loc.gov/2020058666

© 2022 Booklife Publishing
This edition is published by arrangement with Booklife Publishing.

North American adaptations © 2022 Bearport Publishing Company. All rights reserved. No part of this publication may be reproduced in whole or in part, stored in any retrieval system, or transmitted in any form or by any means, electronic, mechanical, photocopying, recording, or otherwise, without written permission from the publisher.

For more information, write to Bearport Publishing, 5357 Penn Avenue South, Minneapolis, MN 55419. Printed in the United States of America.

PHOTO CREDITS

All images are courtesy of Shutterstock.com, unless otherwise specified. With thanks to Getty Images, Thinkstock Photo, and iStockphoto. Background texture throughout - Abstracto. Gravestone throughout - MaryValery. Front Cover - ONYXprj. 4 - pashabo, Lilkin, PixMarket. 5 - Krasovski Dmitri, josefauer, MIKHAIL GRACHIKOV, VikiVector, world of vector. 6 - Dzm1try, CRStocker, Tomacco. 7 - ONYXprj, Robert Spriggs. 8 - Marc Kirouac, Seumas Christie-Johnston, Guppic. 9 - Krasovski Dmitri, chippix, Francis J. Petrie Photograph Collection [Public domain], HappyPictures, artmakerbit. 10 - ShadeDesign, alex74. 11 - Roxana Gonzalez, HappyPictures. 12 - Tymonko Galyna, DbDo, Unitone Vector, zigzag design, Flas100. 13 - Takis Bks, alex74. 14 - BeataGFX, KRPD. 15 - Zentangle, Mark.Ko, Everett Historical. 16 - Mathew Brady [Public domain], E.Thomas (engraver), Augustin Challamel, Desire Lacroix [Public domain], Ysami. 17 - Walters Art Museum [Public domain]*, Everett Historical, NotionPic, riekephotos. 18 - panotthorn phuhual, Rvector. 19 - Rolaks, Glinskaja Olga. 20 - https://commons.wikimedia.org/wiki/File:L%27empereur_Minghuang_regardant_Li_Bai.jpg, Charlesimage, Maike Hildebrandt, blackpencil. 21 - Orest Kiprensky [Public domain], Morphart Creation, A7880S, VectorPlotnikoff, GoodStudio. 22 - asantosg, M Zemek. 23 - Nella. 24 - Agence de presse Meurisse [Public domain]*, Kurilenko Katya. 25 - naulicrea, Everett Historical, Aksanaku, Morphart Creation, NotionPic. 26 - Rachael Arnott. 27 - Danussa, Bonezboy, pashabo, GoodStudio. 28 - Babich Alexander, emka74. 29 - Everett Historical, Laralova, VectorPot. * - U.S. work public domain in the U.S. for unspecified reason but presumably because it was published in the U.S. before 1924.
Additional illustrations by Jasmine Pointer.

CONTENTS

Welcome to the Disaster Zone 4
Bobby Leach 6
Daredevil Days 8
Draco of Athens 10
Fan-tastically Fatal 12
John Sedgwick 14
Famous Last Words 16
Li Bai 18
Going Overboard 20
Franz Reichelt 22
The Flying Tailor 24
Sophie Blanchard 26
The Sky's the Limit 28

Timeline of Deaths 30
Glossary 31
Index 32
Read More 32

WELCOME TO THE DISASTER ZONE

From military generals to lawmakers, from poets and parachutists to balloon pilots and barrel riders, people in the past often led weird and wild lives. So it makes perfect sense that some of those lives ended in ways that were just as strange.

Since the beginning of human history, about 107 billion people have lived on Earth. You know what that means... there are lots of deaths to choose from!

In this book, we are going to look at the stories of six people who were taken out by awful accidents, whether it was inventions that didn't work the way they were supposed to or a surprisingly slippery orange peel.

Into the Disaster Zone We Go . . .

Throughout history, there have been lots of strange sayings that mean someone has died.

Here are a few of the weird ones:

- Kicked the bucket
- Bit the dust
- Met their maker
- Six feet under
- Food for worms
- Pushing up daisies

BOBBY LEACH

Bobby Leach was a **stuntman** who went over the towering Niagara Falls in a barrel in 1911. Bobby used his newfound fame to tour the world with his barrel and share his story. It was on one of these tours that Bobby slipped on an orange peel. He hurt his leg, and the injury became **infected**. Bobby died just two months later.

Bobby was in his fifties when he went over Niagara Falls! The stunt broke his kneecaps and fractured his jaw.

BORN: Cornwall, England
DIED: Auckland, New Zealand
CLAIM TO FAME: Stuntman
DEATH BY: Orange peel

"Would it kill you to pick up your trash?"

1858–1926

DAREDEVIL DAYS

When Bobby was in his sixties, he returned to Niagara Falls—this time to try to swim across its **rapids**. Bobby couldn't manage to do it, and he had to be rescued. He probably should have just retired at this point but instead returned to touring.

The rapids at Niagara Falls

Niagara Falls is nearly 177 feet (54 m) tall and has rocks at its bottom.

STOLEN THUNDER

Before his death by orange peel, Bobby had earned fame, money, and the chance to travel. For Annie Edson Taylor, who was the first person to have successfully gone over the waterfall, things didn't work out so well.

Bobby with his barrel

Annie tested her barrel by sending a cat over the falls in it. Luckily, the cat survived!

Annie hoped her stunt would earn some money. Unfortunately, Annie didn't become famous, and she ended up dying poor.

Annie went over Niagara Falls in 1901, on her 63rd birthday.

9

DRACO OF ATHENS

We don't know much about the ancient Greek **legislator** Draco other than that he was famously mean. Draco is known as the first person in Athens to write laws down—more than 2,500 years ago! The only problem was that his laws were horribly harsh. Surprisingly, Draco still had many fans.

A way of showing appreciation for someone at the time was to shower them with cloaks and hats. Draco's fans apparently did just that. But Draco ended up suffocating under all the clothing they threw at him!

Stealing vegetables was deserving of the death penalty according to Draco!

BORN: Somewhere in Greece
DIED: Aegina, Greece
CLAIM TO FAME: Mean legislator
DEATH BY: Popularity

I think that's enough now, guys!

SEVENTH CENTURY BCE

FAN-TASTICALLY FATAL

Before Draco wrote down his laws, Greece was ruled by only spoken laws. The fact that they weren't written down meant that people could interpret them however they wanted.

Draco's laws were eventually carved into stone tablets.

Draco's written laws meant that everybody had to follow the same rules.

Under **oral** law, people would sometimes just beat each other up if they had a disagreement!

Draco's laws were displayed in places where everyone could see them. This made it so everyone who was able to read would know what the laws were. Draco also introduced the idea of a **court of law** to enforce his rules.

The court was made up of around 400 members, who would meet in places like this theater.

BETTER BUT NOT EQUAL

Draco's rules didn't make things equal. The laws still favored the upper classes. If a lower-class person owed an upper-class person money, the upper-class person could take them as a slave. Yikes!

Themis, Greek goddess of justice, may have had to remove her blindfold to read Draco's laws carefully.

JOHN SEDGWICK

John Sedgwick was a general who fought many battles in the Civil War (1861–1865).

At the Battle of Spotsylvania Court House, John's men ducked to avoid enemy fire. But John stood tall, declaring, "They couldn't hit an elephant at this distance." Just as he spoke, a bullet hit him under his left eye, and he died.

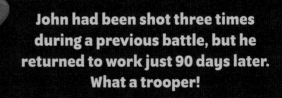

John had been shot three times during a previous battle, but he returned to work just 90 days later. What a trooper!

FAMOUS LAST WORDS

John Sedgwick's last words are famous for how untrue they ended up being. Let's take a look at some other famous last words.

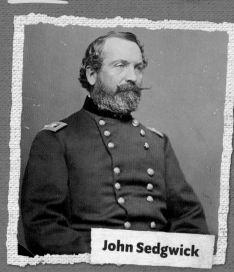

John Sedgwick

Thomas de Mahy

Thomas de Mahy

During the French **Revolution** (1789–1799), the people of France overthrew their king and queen. They sentenced royal supporter Thomas de Mahy to death. On reading his **execution** order, Thomas said, "I see that you have made three spelling mistakes."

"Tomorrow, I shall no longer be here."
— Nostradamus, 1566
This famous **seer** was right—he died the next day.

Saint Lawrence of Rome

In the year 258, Lawrence the **deacon** of Rome, was ordered to bring all of the Catholic Church's money to the Roman emperor. But Lawrence declared that the poor people of Rome were the church's real treasures. The emperor ordered that Lawrence be burnt on a **spit**. During his execution, Lawrence said, "Turn me over; I'm done on this side."

Saint Lawrence's witty taunts of the emperor have made him a patron saint . . . of cooks!

Oscar Wilde

Oscar Wilde said, "This wallpaper and I are fighting a **duel** to the death. Either it goes or I do."

The wallpaper won the duel. Oscar spent his final days surrounded by it in a hotel room in Paris.

Oscar Wilde

LI BAI

Though he wrote in the eighth century, Li Bai is still considered one of the greatest Chinese poets of all time. After becoming a famous writer, Li chose to live like a **nomad** and sail his boat around the country. While wandering around, he may have gotten a little too lonely.

The story goes that Li was sailing down the Yangtze River one night when he saw the moon's reflection in the water. He fell in love with it and leaned over to give it a kiss. In doing so, Li fell out of the boat and drowned.

Even if this story of Li's death isn't true, his travelling lifestyle and poor water safety are probably what killed him.

BORN: Jiangyou, China
DIED: Dangtu, China
CLAIM TO FAME: Poet
DEATH BY: Love for the moon

"I'm coming for you, my love!"

701–762

GOING OVERBOARD

Li Bai's death was just the last in a series of unfortunate events. He'd been fired from a job working for the emperor because he was loud and disrespectful.

The emperor watching as Li Bai writes a poem

Li then joined a **rebellion** against the emperor and was sentenced to death. Luckily, his sentence was reduced to exile.

Li Bai

Exile is when someone is forced to leave a place, usually as a form of punishment that lasts forever.

20

Kiss of Death

Being creative and being a little strange seem to go hand in hand. Let's take a look at some other famous creative people who lived interesting lives.

Alexander Pushkin was a Russian poet who challenged people to duels over the slightest disagreement. He was killed during one of these duels in 1837.

Alexander Pushkin

In the fourth century BCE, the Greek politician Demosthenes would shave half the hair on his head. Why? Afterward, he'd feel too silly to go outside, so he'd stay home and get important work done instead.

Demosthenes

The English poet Lord Byron kept a bear as a pet in the early 1800s.

21

FRANZ REICHELT

Franz Reichelt was a tailor with a dream. In 1910, he invented a parachute that was also a coat. Two years later, he wanted to show it off where everyone could see. And what better way to do that than jumping off the Eiffel Tower?

Franz was so confident in his invention that he wanted to make the leap himself. As the crowd watched from below, Franz jumped off the tower. Unfortunately, his coat did not work.

Franz didn't know it, but an American named Frederick R. Law had successfully parachuted off the Statue of Liberty two days earlier.

THE FLYING TAILOR

Franz would attach early versions of his coat to dummies and throw them off the roof of his shop. Some dummies fell slowly enough to land safely.

Franz Reichelt, wearing his parachute-coat

Franz began to test the coat himself, and despite badly hurting his leg, he still believed in his invention.

Franz Reichelt, just before his jump

Franz was convinced that he needed to jump from a greater height to test the parachute properly.

Believe In Yourself

Way back in the eleventh century, Abu Nasr Ismail ibn Hammad al-Jawhari shared Franz's passion for flying—and his sad end. He launched himself from the roof of a **mosque** with wooden wings strapped to his arms. Like Franz's coat, they did not work, and Abu fell to his death.

A Foot in the Grave

William Bullock died in 1867 after he got his foot trapped in a printing press he had invented.

Printing press

Flying High

Otto Lilienthal, the inventor of the hang glider, died in 1896 during a test flight.

Otto Lilienthal

SOPHIE BLANCHARD

In 1805, Marie Madeleine-Sophie Armant became the first woman to pilot her own hot air balloon as well as the first woman to be a professional **balloonist**. People found her really interesting, and they wanted to see her fly her balloon in **exhibitions**.

Sophie's downfall came during one of these exhibitions. She lit fireworks inside her hot air balloon, which caused it to catch fire and crash into a house. The balloon was damaged beyond repair and—sadly—so was poor Sophie.

Sophie became known as Madame Blanchard after she married her husband, Jean-Pierre Blanchard.

THE SKY'S THE LIMIT

The love that pioneering female **aviators** like Sophie had for flying inspired other women to take to the skies.

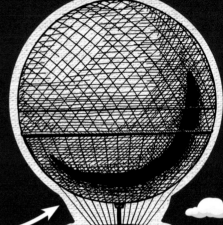

An illustration of a balloon belonging to Jean-Pierre Blanchard, Sophie's husband

Harriet Quimby

In 1911, Harriet Quimby became the first American woman to get her pilot's license. The following year, she became the first woman to fly solo across the English Channel.

As an African American, Bessie Coleman was not allowed to attend flight schools in the United States. In 1920, she moved to France to follow her dream, becoming the first African American to get a pilot's license.

Bessie Coleman

Amelia Earhart holds the title of first woman to fly over the Atlantic Ocean on two occasions—first as a passenger in 1928 and then solo in 1932.

Amelia Earhart was flying around the world in 1937 when she and her co-pilot went missing over the Atlantic Ocean.

Amelia Earhart

TIMELINE OF DEATHS

GLOSSARY

aviators people who fly aircraft

balloonist a person who flies a balloon

court of law a group of people brought together to settle criminal disputes

deacon an official in a Christian church who has special duties and jobs to perform

duel a fight or argument between two people, sometimes involving weapons

execution the act of killing someone as punishment for a crime

exhibitions events where objects, art, or performances are shown to the public

infected when a wound or body part is diseased because bacteria or a virus got inside it

legislator a person who makes laws

mosque a building that is used for Muslim religious services

nomad someone who constantly moves from place to place and has no permanent home

oral spoken rather than written

rapids areas of shallow, fast-moving water

rebellion an effort by many people to change a government or leader of a country by protest or force

revolution the attempt by many people to overthrow a government and start a new one

seer someone who predicts things that will happen in the future

spit a thin, pointed rod for holding meat over a fire

stuntman a man who performs difficult physical actions that require lots of skill and courage

INDEX

balloons 4, 26–28
barrels 4, 6, 9
duels 17, 21
dummies 24
emperors 17, 20
executions 16–17
flying 24–26, 28–29
France 16, 23, 27, 29
injury 6
inventions 5, 22, 24–25
inventors 23, 25
laws 10, 12–13
Niagara Falls 6, 8–9
parachutes 22, 24
pilots 4, 26, 28–29
poets 4, 18–19, 21
revolution 16
stunts 6, 9
war 14

READ MORE

Finan, Catherine C. *Ancient Greece (X-treme Facts: Ancient History).* Minneapolis: Bearport Publishing, 2022.

Harris, Tim. *Triumphs of Human Flight: From Wingsuits to Parachutes (Feats of Flight).* Minneapolis: Hungry Tomato, 2018.

Smith, Matthew Clark. *Lighter Than Air: Sophie Blanchard, the First Woman Pilot.* Somerville, MA: Candlewick Press, 2017.